Downtown Details

AN ARTFUL PERSPECTIVE OF NEWPORT, PENNSYLVANIA

PERRY COUNTY COUNCIL OF THE ARTS

an imprint of Sunbury Press, Inc.
Mechanicsburg, PA USA

an imprint of Sunbury Press, Inc.
Mechanicsburg, PA USA

For information about special discounts for bulk purchases, please contact Sunbury Press Orders Dept. at (855) 338-8359 or orders@sunburypress.com.

To request one of our authors for speaking engagements or book signings, please contact Sunbury Press Publicity Dept. at publicity@sunburypress.com.

ISBN: 978-1-62006-761-1 (Trade paperback)

Library of Congress Control Number: 2018941967

FIRST BROWN POSEY PRESS EDITION: April 2018

Product of the United States of America
0 1 1 2 3 5 8 13 21 34 55

Set in Bookman Old Style
Designed by Crystal Devine
Cover by Lawrence Knorr
Cover photo by Irene VanBuskirk
Edited by Lawrence Knorr

Continue the Enlightenment!

Artful photos by Irene VanBuskirk
Document photos by Jane Hoover
Historic photos courtesy of the Perry Heritage Collection
and Louise Beard Museum

Contents

Foreword

"I never realized Newport had such marvelous architecture until Andrea told me to look up!" Thus, spoke community leader Frank Campbell after Andrea MacDonald, State Historic Preservation Officer, escorted a group of us around Newport in 2017. Now, through the lens of artistic photographer Irene VanBuskirk and the historic and architectural narratives of writers Jane Hoover and Roger Smith, we too may tour Victorian-era Newport.

Newport has always been a transportation hub for the region, first by river, then canal and railroad, and now by highway. Most of the structures featured in this delightful and well-composed and designed volume were built in the economic boom times of the era, from the 1870s to the 1920s. Greater Newport workers produced iron, leather, grains, flour, and lumber and constructed numerous factories and warehouses before loading goods on the two railroads that the served the community – the Pennsylvania Railroad and the Newport and Sherman's Valley Railroad.

Merchants, teachers, lawyers, doctors, and laborers were needed to support the high level of production, and all required housing and buildings for their services. Contractors and builders remained busy for two generations constructing the residences and businesses that remain in Newport to this day. Fortunately for us, many of the buildings retain their original materials and architectural features. Others have been modified and updated, but it is the ones that have not been touched that reveal the most suprises. And yes, by looking up, we too can see the craftsmanship and artistry of that period, still on display for this generation and those to come.

—Glenn N. Holliman, Chair,
Perry County Bicentennial Committee

Acknowledgments

This project was supported in part by the Pennsylvania Council on the Arts, a state agency funded by the Commonwealth of Pennsylvania and the National Endowment for the Arts, a federal agency. Historical photographs were supplied courtesy of the Perry Heritage Collection and the Louise Beard Memorial. A special thank you to the owners of the buildings featured. Without the assistance of the following, this exhibit would not have been possible: Jasmine Colbert, Perry County Council of the Arts, graphic designer; Toni Semanko, photo editor; Jerry Clouse, MA, architectural consultant; Jeffery Lynn Wright, history consultant; Kay Cramer, grammar editor; Cyndi Long, GFADesign.com, graphic designer; Andrea MacDonald, State Historic Preservation Officer and Bryan Van Sweden, Community Preservation Coordinator of the Susquehanna Region. Thank you also to all the residents of Newport and surrounding areas who have shared their records and stories, including members of the Perry Historians, Edna H. and Elmer R. "Bob" Baker, Jr., Suzanne B. Beamer, Jerry A. Clouse, Robert M. Flickinger, Harry A. Focht, Jean H. Tuzinski, and Jeffery Lynn Wright.

The authors would also like to thank family and friends, who supported us as we focused on this project for many weeks.

Newport Historical District Map

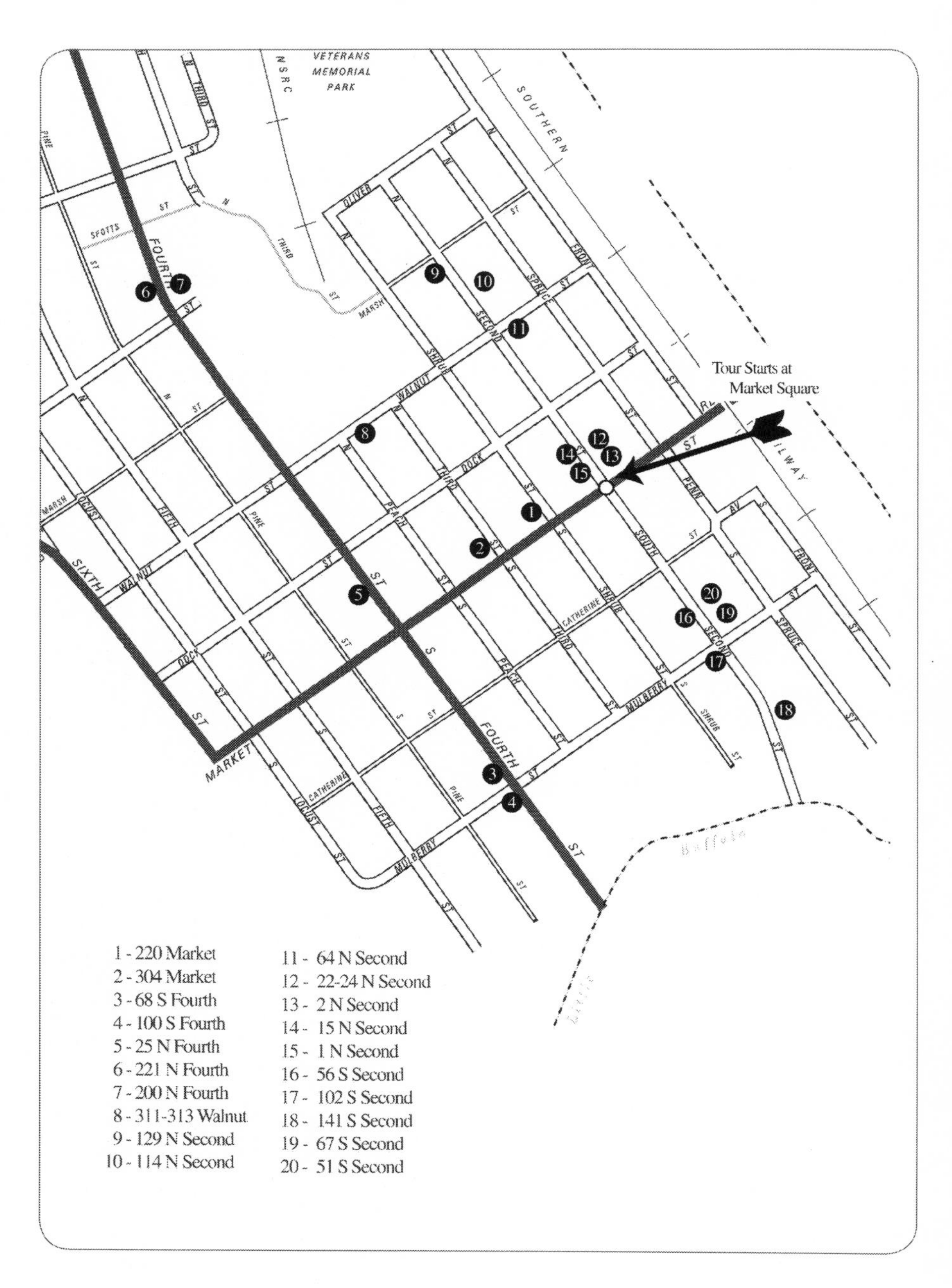

Introduction

MORE THAN A HUNDRED YEARS AGO, anonymous Victorian craftsmen created beautiful details for the buildings of downtown Newport. These details were the manifestation of their skill and creativity, the prosperity of the borough, and the architectural movements of the era.

Many people walk by these buildings every day and never look up to notice the details. Or don't know what they are called. Or don't think about why they look that way.

It is our desire to inspire you to look up and to appreciate these imaginative architectural details by showing them to you through the eye of a photographer, the research of an amateur historian and an amateur architect:

Irene S. VanBuskirk • Jane T. Hoover • Roger A. Smith

We also hope that, on your next walk down the streets of Newport, you will even find details that we have missed. Look Up!

Despite three floods and periodic economic downturns, magnificent examples of Victorian architecture survive in Downtown Newport. The term "Victorian" refers to the era during which a building was constructed—during the reign of Queen Victoria of England [1837-1901]—not its style of architecture. The period coincided with relative economic prosperity in Newport. Newport was a stop on the Pennsylvania Canal and later the Pennsylvania and Newport and Sherman's Valley railroads and thus a commercial hub in the area. Local foundries, brickyards, and planing mills were sizeable employers. They were also a source of relatively inexpensive materials for many of these buildings. Styles during this period included Queen Anne, Carpenter Gothic, Second Empire, Classical Revival, Picturesque, Georgian Revival, and Italianate. Most Newport residents were not purists. They selected elements from styles that they liked. During this romantic period, people's taste dictated that fashion and furnishings, as well as architecture, should be beautiful and ornate, rather than practical.

Although styles of Victorian architecture varied widely, there were commonalities:

- Structures were usually large and imposing, frequently of two or three stories. Large, one-story porches were popular.
- Exteriors were made of wood, stone, or brick.
- Buildings had complicated, asymmetrical shapes.
- Trim was elaborate and frequently ornamental rather than practical.
- Exteriors made lavish use of textures such as scalloped shingles, patterned masonry, or siding of different widths.
- The height of the buildings might be accentuated through such vertical elements as steep roof lines and towers.
- Siding and trim were often painted in vibrant, complementary colors.

Beyond the eclectic architectural styles, Newport houses incorporated significant innovation during the Victorian era:

- Sanitation: Toilet facilities moved from backyard privies to indoor toilets.
- Water: Access moved from a backyard pump to running tap water. Later, water warmed by a boiler fed both kitchen and interior bathtubs.
- Lighting: Interiors were illuminated first by candles, then by oil lamps, then by gas jets, and finally by electricity.
- Basements: Central heating became possible with coal-fired furnaces in the basement.
- Windows: Larger panes of glass became possible as glass-making technology improved.

1

220 Market Street

ORNATE ELEGANCE

[A] **Palladian windows**—a three-section window where the center section is arched and larger than the two side sections. Palladian windows, elements of a style dating back to the Renaissance era, are designed to let in additional light and to create a more stately and formal feeling.

[B] **Cornice**—a roof eave finished with a decorative molding. Eaves, or roof overhangs, are functional. Their purpose is to direct rainwater away from a building's walls and foundation and to shade the building from sunlight during the summer.

[C] **Bracket**—a structural or decorative element that supports an eave or cornice. Although brackets appear to support an eave or overhang, their presence may be merely decorative. Brackets can be made of wood, stone, plaster, or metal.

[D] **Frieze**—a broad horizontal band of sculpted or painted decoration. Note the frieze windows placed under the cornice and between the brackets.

Italianate

Italianate houses are easily distinguished by their gently sloping roofs and deep overhanging eaves, which are seemingly supported by a row of decorative brackets or modillions. The exterior of Italianate houses can be made of wood, brick, stone, or stucco. From the late 1840s to 1890, Italianate architecture achieved huge popularity in the United States.

This Italianate house is side-gabled, built of brick on a stone foundation. It has a large Roman-arched entrance with a fanlight, wooden lintels over the windows, and teardrop eave brackets. The brick looks handmade, with very thin mortar. The sandstone porch with its Ionic columns was built around 1915.

The multiple colors on this building probably reflected the original. Two or three colors were common during the Victorian period. Note the Ionic columns and arched door.

Modern Conveniences

On March 26, 1869, Albert Demaree, a Newport businessman, conveyed two lots with a frontage of 140 feet on Market Street and a depth of 140 feet along Third (Railroad) Street to David L. Tressler, who in turn sold it to William Heim Minnick and his wife Sarah on August 4, 1870. Minnick was a teacher at Loysville Academy, which became Tressler's Orphans Home. Sarah's mother lived in this house with them in her last years and was rescued by boat from the stairs to the second floor during the Johnstown Flood of 1889. William's will, dated June 17, 1907, awarded the house to his daughter Emma. Emma married her cousin Arthur Tressler Scott, who was a professional singer. She played the piano and violin and assisted in her husband's musical work.

On January 16, 1912, Emma and Arthur conveyed the property to Dr. Sam Whitmer, DDS. Dr. Whitmer replaced the small front stoop with a large porch with Ionic columns and changed one of the large windows to a door for the entrance to his dental office. He installed the first steam heat furnace in Newport and added one bathroom, upstairs. Dr. Whitmer died April 28, 1926, leaving the property to Belle Ballard Whitmer. In 1927, Belle began renting rooms and office space to Dr. Edward P. Hewlings, DDS, and to Dr. Blaine F. Barto, MD. In 1938 Belle divided the property, selling the dwelling and a lot with 60 feet of frontage on Market Street to Dr. Hewlings and a vacant lot immediately west of the house to Dr. Barto. Dr. Hewlings used the house as his residence and office until his death on April 17, 1968. The widow's walk was removed during Dr. Hewlings' ownership.

220 Market Street

2

304 Market Street

Reserved Elegance

[A] Lintel—a structural horizontal block that spans the space or opening in doors, windows, and fireplaces. Lintels can have a structural purpose, can be merely a decorative element, or can be both.

[B] Sill—a shelf or slab of stone, wood, or metal at the foot of a window or doorway.

Georgian Revival

Georgian Revival architecture is marked by symmetry and proportion based on the classical architecture of Greece and Rome. Ornamentation is typically restrained, and sometimes almost completely absent on the exterior. As is often the case in American architecture, this plan is a "mix and match," with an Italianate doorway, characterized by its arch and a reserved but three-dimensional construction.

Circa 1890

This large two-story side-gabled brick house sits on a stone foundation. It has five bays, a Roman-arched entrance with a transom, wooden lintels over the windows, faux shutters, and a small one-story porch with square wooden columns.

The sills on this building are smaller than the lintels and understated. They are the same color as the shutters adding another textural element to the façade. The snow guards on the roof have a practical purpose. They prevent large masses of snow or ice from descending from the roof all at once.

A Writer's Nook

John B. McAllister, W. H. McAllister, and his wife Rebecca of Fayette Township purchased the lot from Joseph Tate, who developed Newport westward from Second to Fourth Street. In 1861, they sold it to Joshua S. Leiby, dry goods merchant, and his wife Susan for $437.

The 1863 map of Newport shows a large house owned by J. Leiby at this location. However, the floor joists, as well as a Third Street basement window, show a date of October 10, 1871, initialed BFD. The header of the basement staircase has the initials JFD, with no date.

On October 1, 1908, Harry H. and Frances Hain of Newport Borough purchased the house from Scott Leiby, executor of Joshua's will, for $3,700. Harry was the great-grandson of Jacob Huggins, founder of Liverpool, county commissioner in 1820, and member of the General Assembly in 1824. Hain worked on his monumental *History of Perry County, Pennsylvania* while living in this house.

However, the book was not published until 1922, nine years after he left Newport to live in Harrisburg.

Hiram Martin Keen of Tyrone Township and his wife Hannah Groff of Quarryville purchased the house for $4,900 in 1913. When Hiram died in 1919, ownership was held equally by Hannah & their only son James, who conveyed his share to his mother June 21, 1920, for $3,000. Mrs. Keen lived here until 1953 when Roy Miller Duffy of Hollidaysburg and his wife Mary Flurry bought the house for $8,100.

304 Market Street

3

68 South Fourth Street

A Prominent Mansion

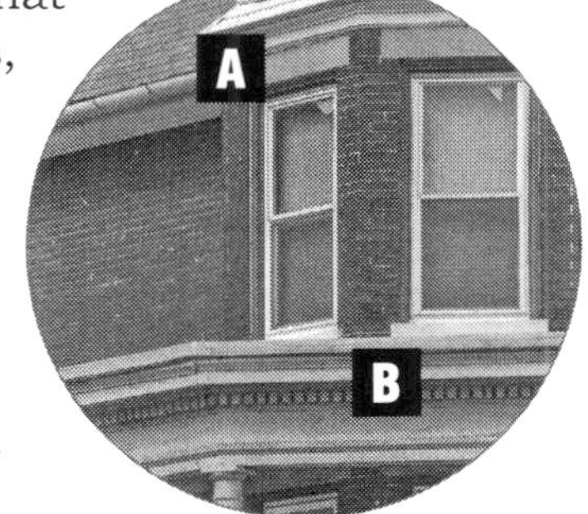

[A] **Lintels**—a structural horizontal block that spans the space or opening above doors, windows, and fireplaces. Lintels can have a structural purpose, can be merely a decorative element, or can be both.

[B] **Dentil**—a small block used as a repeating ornament in a cornice. Dentils are found in classical Greek and Roman architecture and in later eclectic styles.

[C] **Baluster**—a short pillar or column, typically decorative in design, placed in a series to support a handrail. Balusters have several purposes within the overall structure: support the handrail, provide safety, and provide additional style and structural flair.

QUEEN ANNE

The house reveals several Queen Anne characteristics including a steeply pitched roof with a lower cross gable to the right of the tower and a shed-roof dormer to the left, giving it an asymmetrical appearance. The large wrap-around porch is supported by Tuscan columns.

Note the various textures in this building, the use of different colors, and how the large porch draws people outside.

OLDEST HOUSE IN NEWPORT REPLACED

The News of January 14, 1905, reported:

> The oldest house in Newport was razed on Monday. It was built in 1839 by Joseph Tate, and for the past thirty-two years has been in possession of ex-Associate Judge James Everhart, now deceased.

Joseph Tate was a key figure in Newport history, having expanded Paul, John, and Daniel Reider's town plan from 54 lots to 145 in 1834. The old house sat on the northern side of the lot, along what is now Fourth Street. The house and lot passed through several hands, returning to Joseph R. Tate and his wife Annie E. (Bortel) in 1867. In November 1867, they conveyed the house and lot to Emory S. Bortel. In April 1873, Judge James Everhart bought the house and lot. James Everhart's will granted the lot to his daughter Anna Belle and six months later she tore the old house down.

The large Queen Anne house we see today was built between 1905 and 1911. Anna Belle Everhart died in 1920 and the house passed to Robert W. Clark.

In 1926, Harry H. Deckard bought the house at a Sheriff's sale and in 1967 his widow Hattie L. Deckard conveyed the house to her daughter Isabel Deckard Donaghy and her husband Thomas J. Donaghy, publishers of *The News-Sun*.

68 South Fourth Street

4

100 South Fourth Street

Newport's Show Place

[A] Palladian Window—a three-section window where the center section is arched and larger than the two side sections. Palladian windows, elements of a style dating back to the Renaissance era, are designed to let in additional light and to create a more stately and formal feeling.

[B] Cornice—a roof eave finished with a decorative molding. Eaves, or roof overhangs, are functional. Their purpose is to throw rainwater away from a building's walls and foundation and to shade the building from sunlight during the summer.

[C] Modillions—an ornament that supports the cornice. In modern architecture, modillions are decorative, not functional.

[D] Quoins—installed on the corner of a building to call attention to the corner and give it the appearance of permanence and stability.

[E] Column—a structural element providing support. The top of a column, or capital, comes in numerous styles. This one is called Ionic.

HIPPED ROOF

The historical picture below displays a good example of a hipped roof, where all four sides slope down to the walls, with a gentle slope. Notice also the beautiful example of a widow's walk.

This large, two-and-a-half-story house is known as the J. Emery Fleisher mansion. It is five bays wide and four deep with a Palladian window with stained glass on the second floor, bay windows on each side, an elaborate entrance door flanked by Doric columns, stone lintels, and window sills, a porte-cochère (covered carriage entrance), quoins, dormers, and a one-story full front porch supported by clustered Ionic columns set on brick piers.

ALL IN THE FAMILY

In 1877, a Mrs. Leiby owned a small house on this site, at the corner of Fourth and Mulberry Streets. J. Harry McCulloch sold the property to his step-father, J. Emery Fleisher, in August 1901. The Fleisher Mansion was built by 1902, according to the records of the Sanborn fire insurance maps. Upon his death, Fleisher deeded the property to his three stepsons J. Harry McCulloch, DDS, William Ross McCulloch, and David H. McCulloch. In 1935, J. Harry

McCulloch conveyed the property to Charles McHenry Eby, son of Dr. James Eby, a prominent Newport physician in the late 19th century. In 2018, the house is used as an apartment building.

100 South Fourth Street

5

25 North Fourth Street

BUILT IN 1912

[A] **Quoins**—installed on the corner of a building to call attention to the corner and give it the appearance of permanence and stability.

[B] **Tower**—an architectural element that is significantly taller than it is wide and that originates at ground level.

[C] **Corbel**—projection of brick jutting out from a wall to support a structure above it. The corbels are more obvious in the 1912 schoolhouse photo on the next page.

[D] **Arched Doorway**—indicates Roman Revival rather than Greek Revival, which would have had a horizontal doorway.

[E] **Sandstone Sill**—provides visual contrast to red stone and brick.

Victorian Romanesque

This two-story brick building was constructed on a contrasting red stone foundation with a Roman-arched door and windows in the central portion and large, one-bay-wide towers flanking the entrance. It is now an apartment building.

Enrollment

By 1898, Newport had outgrown the original school on this site and classes were dispersed around town while a new school was built. The new school was completed in November 1912, providing space for all twelve grades and 304 students. When Oliver Township's Evergreen School was destroyed by fire in 1926, Oliver and Newport School Districts joined, forming the Newport Union School District. Within a year or two, the remaining Oliver Township schools had closed, doubling the enrollment in Newport schools. In 1928, the School Board opened a new high school at Fifth and Caroline Streets.

Former 1866 schoolhouse at 25 North Fourth Street

1912 schoolhouse

25 North Fourth Street

6

221 North Fourth Street

PRESBYTERIAN MANSE

[A] **Tower**—an architectural element that is significantly taller than it is wide and that originates at ground level.

[B] **Baluster**—a short pillar or column, typically decorative in design, placed in a series to support a handrail. Balusters have several purposes within the overall structure: support the handrail, provide safety, and provide additional style and structural flair.

[C] **Spindlework**—delicately turned supports, used as a frieze suspended from the porch ceiling.

[D] **Sidelight**—a narrow window or pane of glass set alongside a door or larger window.

RESURRECTED BEAUTY

This three-story house is side-gabled with two bays. A unique feature is the diagonally-projecting gable end tower on the southeast corner. The house was clad in vinyl around 2000 and the columns, balusters, and spindles were replaced with vinyl. However, the front door retains its original sidelights. Notice how the choice of architectural elements and dramatic colors retain the Victorian character of the house.

CARPENTER'S MARK

In 1886, North Fourth Street was extended from Oliver Street to Fairground Road, there joining Front Street and proceeding on across the Big Buffalo Creek. Fairground Road was then known as the road to Milford (now Wila). D. H. Spotts, having inherited the land from his step-father Jesse L. Gantt, sold this lot and several others just north of Oliver Street in the 1890s. In 1897, the residents of Spotts' development petitioned the Court of Quarter Sessions to become part of the Borough of Newport rather than of Oliver Township. Accordingly, the court decreed that the land from Oliver to Fickes streets and the River to Sixth Street become part of the Borough.

Spotts sold this lot along North Fourth Street to Edwin and Minnie Morrow in 1896. A "carpenter's mark" on the underside of the drawer located beneath the back stairway leading from the kitchen to the second floor confirms that the house was completed on December 16, 1896. The Morrows sold the property to the Newport Presbyterian Church on March 22, 1902.

The Rev. Harry Milton Vogelsonger's daughter Helen Mae and Pauline Whitekettle from 253 North Fourth played together on the third floor of the tower during the 1920s. The church used it for a manse until 1946 when it was sold to Herbert and Lucy Farnsler. A year later, Luther J. and Virginia Kell Mattern bought the house and it has remained in the Mattern family ever since.

221 North Fourth Street

7

200 North Fourth Street

BUILT CIRCA 1890

[A] **Drip Cap**—the horizontal molding above the top edge of lintels, windows, and doors. The purpose of a drip cap is to direct water away from the window and directly to the ground.

[B] **Finial**—a decorative ornament placed at the top or end of a roof, gable, or pole.

[C] **Sunburst**—an architectural ornament that consists of rays or “beams” radiating out from a central disk to evoke sunbeams. Sunbursts can be circular, semicircular or semi-elliptical in shape.

[D] **Gable**—the triangular upper part of a wall at the end of a ridged roof.

CARPENTER GOTHIC

Carpenter Gothic houses became common in North America in the late nineteenth century. The invention of the scroll saw and mass-produced wood moldings made it possible for middle-class people to embellish their homes relatively cheaply.

The major architectural elements of the house are symmetrical. The front door is centered between two large, forward projecting bay windows rising to gable-end dormers. The fan decoration over the windows and the sunbursts and vertical boards in the gables provide texture and draw the eyes up.

THE DOCTOR IS IN

In 1889, David H. Spotts, an early Newport developer, sold the land to Charles L. and Kate VanNewkirk who probably built this house around 1890. They, in turn, sold the property to Horace B. Light in 1900. It remained in the Light family until 1946 when Dr. Leonard Beaver Ulsh, MD, bought it.

Dr. Ulsh graduated from Newport High School in 1922 and from the University of Pennsylvania Medical School in 1930. In 1932, the Presbyterian Board of Foreign Missions commissioned him to be a medical missionary to Khartoum, North Sudan. After six months in London studying tropical medicine and a short time in Cairo learning the language, he and his wife traveled to North Sudan to begin their work. They returned to Newport where Dr. Ulsh practiced medicine in Newport for many years, dying in 1970.

200 North Fourth Street

8

311-313 Walnut Street

SECOND EMPIRE

[A] **Mansard Roof**—a roof with two distinct slopes, the upper one being shallow and the lower one being steep. Mansard roofs are usually characterized as four-sided, but in town, houses are often only two-sided.

[B] **Hooded Dormer**—a window set vertically into a roof. These are topped with gables resembling hoods serving to keep the rain off the windows.

[C] **Finial**—a decorative ornament placed at the top or end of a roof, gable, or pole.

[D] **Bracket**—a structural or decorative element that supports an eave or cornice. Although brackets appear to support an eave or overhang, their presence may be merely decorative. Brackets can be made of wood, stone, plaster, or metal. These end brackets are enlarged and are topped with gabled finials.

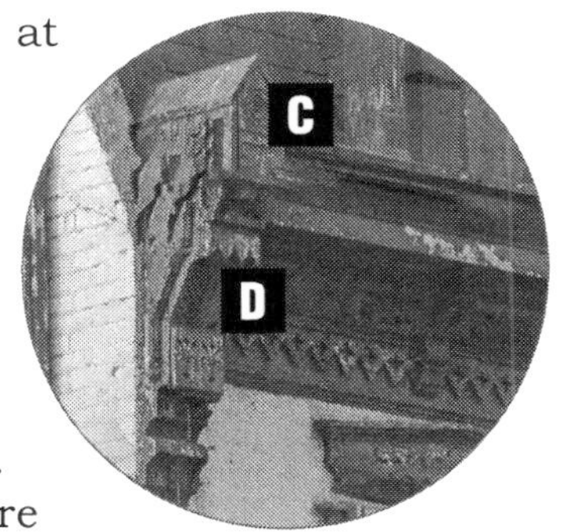

DETAIL

Because of its roof, this building can be termed Second Empire Style, which is characterized by mansard roofs, elaborate dormers, and massive façades. At the peak of its popularity in the United States (roughly 1855-1885), the style was considered both fashionable and a statement of modernity. Its popularity led to a widespread remodeling boom during which mansard roofs were incorporated into formerly pitched-roof residences.

THE HISTORY

On March 7, 1884, Augustus and Louise Rippman of Philadelphia sold the property to Charles W. Smith for $650. Smith operated the Adams Express Agency, first

House with a gabled roof, circa 1885

House with a Mansard roof, circa 1900

from his house, then from across Walnut Street in what is now the Weis Market parking lot. Adams Express shipped everything from documents and parcels to crates of chickens. *The News* of November 21, 1895, reported that the Pennsylvania Railroad train known as the *Atlantic Express,* while speeding down Third Street, hit a truck full of chicken crates near the Express office, scattering chickens throughout the freight yard. It narrowly missed Charles Smith and his assistant Harry Flickinger.

Smith drastically remodeled the house into its Second Empire style in July of 1892. The current porch is modern.

311-313 Walnut Street

9

129 North Second Street

Louise Beard House

[A] Gable—the triangular upper part of a wall at the end of a ridged roof. Note the porch roof also has a gable.

[B] Bracket—a structural or decorative element that supports an eave or cornice. Although brackets appear to support an eave or overhang, their presence may be merely decorative. Brackets can be made of wood, stone, plaster, or metal.

Carpenter Gothic

The Carpenter Gothic style became common in North America in the late nineteenth century. The invention of the scroll saw and mass-produced wood moldings made it possible for middle-class people to embellish their homes relatively cheaply.

This clapboard house displays a complex design and the elaborate decoration common to the Carpenter Gothic style. Note the use of texture and color to highlight the verticality of the house.

Texture is the organization of architectural elements such as windows, doors, solids, or voids to create different patterns. There are many textures in this house: horizontal siding, scrollwork under the balcony, and vertical details under the gable.

Color, like texture, is a way to call attention to various architectural elements and please the eye.

Features accentuating the verticality of the house include:

- The clapboards on the second floor are narrower than on the first floor.
- The roof gable is steeper than the porch gable.
- The vertical details and colors under the roof gable draw your eyes up.

Built around 1893

Horace Beard bought the property at public auction for $796. He demolished the old house and built the one we see today. Beard was the nephew of Harry Bechtel, owner of the Bechtel Tannery. Bechtel sold the tannery to the Elk Tanning Company in 1893, and Beard became superintendent of the tannery from 1893 to 1911. In 1938, Katherine Beard bequeathed the house, to be known as the Louise Beard Memorial, to the town in memory of her daughter Louise who died at age 42.

The Beard Family

129 North Second Street

10

114 North Second Street

Decorative Trim

[A] Dormer—a window that projects vertically from a sloping roof.

[B] Cornice—a roof eave finished with a decorative molding. Eaves, or roof overhangs, are functional. Their purpose is to throw rainwater away from a building's walls and foundation and to shade the building from sunlight during the summer.

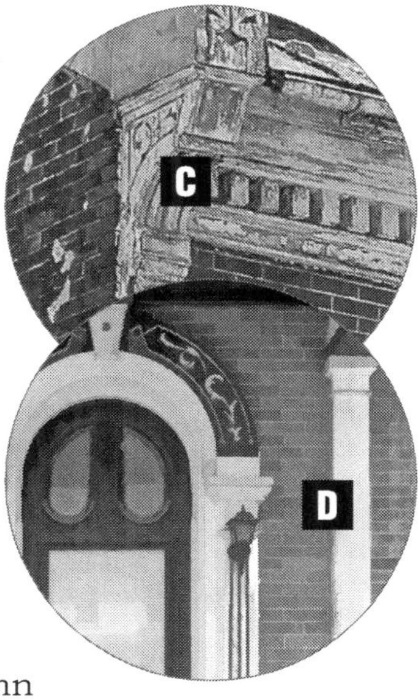

[C] **Bracket**—a structural or decorative element that supports an eave or cornice. Although brackets appear to support an eave or overhang, their presence may be merely decorative. Brackets can be made of wood, stone, plaster, or metal. These end brackets are enlarged and are topped with gabled finials.

[D] **Pilaster**—an architectural element used to give the appearance of a supporting column and to articulate the extent of a wall, with only an ornamental function. It consists of a flat or molded surface raised from the main wall surface, usually treated as though it were a column with a capital at the top and a base at the bottom. In contrast to a pilaster, an engaged column supports the structure of a wall and roof above.

VICTORIAN

This two-story side-gabled brick house is built on a stone foundation. Note the Roman-arched door, the pediments with scrollwork over the windows and door, and the elaborately detailed dormers and eave brackets.

DETAILED CRAFTSMANSHIP

Around 1840, John Beatty built a house here. It would have been one of the first houses north of Center Square. In 1878, Rachel E. Campbell sold the property to George Fleisher, who at that time was part owner of the Newport Planing Mill Company. Judging from the detailing and fine millwork we can assume that Fleisher was responsible for building this house. Perhaps the house was even meant as a marketing tool for the products the Planing Mill could provide. In 1880, he conveyed the property to P. K. Brandt, cashier of the Peoples Bank, reorganized and renamed the First National Bank on July 1, 1893. Brandt remained here until 1900, after which the house passed first to the Swab family and then several others.

114 North Second Street

11

64 North Second Street

Simple Elegance

[A] Dormer—a window that projects vertically from a sloping roof.

[B] Lintel—a structural horizontal block that spans the space or opening above doors, windows, and fireplaces. Lintels can have a structural purpose, can be merely a decorative element, or can be both.

[C] Sill—a shelf or slab of stone, wood, or metal at the foot of a window or doorway.

[D] **Keystone**—a central stone at the summit of an arch, locking the whole together.

[E] **Arch**—a curved symmetrical structure spanning an opening and typically supporting the weight of a wall or roof.

[F] **Bay window**—a window or set of windows that project out from an exterior wall.

D
E
F

Contrast

A large, one-story full front porch, an arched stained-glass window, and a bay window adorn this two-story, tan brick house.

Note how the smooth brick contrasts with the rustic stone lintels and sills and how the major architectural elements in the front of the house are asymmetrical: two windows to the right of the door, one to the left.

Working from Home

W. H. Musser bought the lot from Benjamin A. and Eliza Jenkins in 1857. At that time, there was a frame house, built around 1835, on the southern end of the lot. Samuel Myers acquired the lot from W. H. Musser in 1907 and moved the frame house to Walnut Street and Rose Alley. He then built the present house and turned the frame building into the funeral parlor. He opened a furniture store next to his house on September 16, 1909. When Samuel died in 1941, David M. Myers lived here and operated the businesses.

Frame house now the funeral parlor

64 North Second Street

12

22-24 North Second Street

Italianate Beauty

[A] Low Pitched Roof—a roof with relatively gentle slope from the peak to the eaves.

[B] Drip Cap—the horizontal molding above the top edge of lintels, windows, and doors. The purpose of a drip cap is to direct water away from the window and directly to the ground.

[C] Sill—a shelf or slab of wood, stone or metal at the foot of a window or doorway.

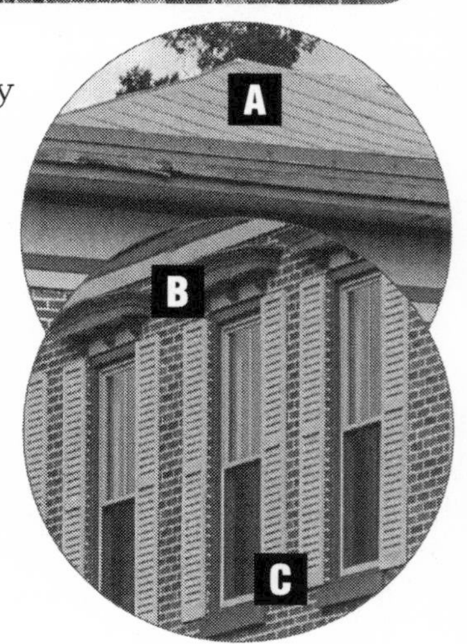

Italianate

1889, with eave brackets

Italianate houses are easily distinguished by their gently sloping roofs and deep overhanging eaves, which are often supported by a row of decorative brackets or modillions. The exterior of Italianate houses can be made of wood, brick, stone, or stucco. From the late 1840s to 1890, Italianate architecture achieved huge popularity in the United States.

This three-story building has a gently sloping roof with a wide overhang, but the original eave brackets visible in the historical picture have been removed. There are Italianate pediments or drip caps over the windows. The storefront windows were removed around 1985 when the building was transformed into office space.

Spared

Dr. Jacob M. Miller and his brother George constructed the building in 1867, probably for W. A. Sponsler and B. F. Junkin, bankers in New Bloomfield. In 1872, they sold it to Caroline Sheibly Spotts Gantt and it later became the residence of her son David H. Spotts. When the 1874 fire consumed the block north of Market Street, from Front to Second, it spared this building because David Spotts had hung water-soaked

Circa 1940, with storefront windows

carpets from the storeroom roof. In 1913, Noll's Café occupied the southwest corner of the building. The poster advertises *The Cave Dweller's Romance*, then playing at the Pastime Theater on Walnut Street where the Post Office is today. David's widow Emma sold the building to Henry Lipsitt in 1919, who in turn sold it to J. C. Bistline in 1926. By 1931 it was the property of Samuel Zuckerman who operated a clothing store here for years.

Storefront

22-24 North Second Street

13

2 North Second Street

A Prestigious Vantage Point

[A] Dentil—a small block used as a repeating ornament in a cornice. Dentils are found in classical Greek and Roman architecture and in later eclectic styles.

"M" is for Mingle House

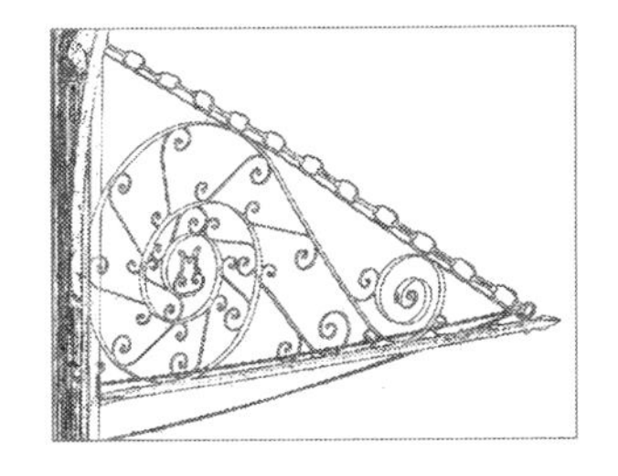

This is a large three-story brick building with a cross-hipped roof, built in 1874. Attached to the Market Street side is a second three-story brick on stone foundation structure, also built in 1874.

Notice the arched bricks acting as lintels.

Luxurious Accommodations

A blacksmith shop is the first recorded occupant of this site. Samuel Sipe built a hotel and tavern here in 1831, and John T. Robinson became the owner from 1837 until 1850 when Jesse L. Gantt bought the property. A fire on June 25, 1874, started in the stable behind the hotel and destroyed the original building, but Gantt rebuilt it immediately. Apparently, the hotel was luxurious for its time. An article in *The News* from November 21, 1874, included the following description:

> On the second and third floors, each, there is a tank with a capacity of 800 gallons each; these are ordinarily supplied with water from the roof, thus affording soft rainwater; but when this fails in long dry weather, they can be filled in a few hours from a force pump in the rear part of the building. Pipes pass from these tanks to the various rooms; there are also pipes containing hot water, so that cold and hot may be drawn at pleasure. The water is heated by the range and passes off in the pipes to the several parts of the building. An appropriate receiver is also fixed on each story into which all wastewater can be thrown, from whence it is carried into the main sewer by means of pipes. No water need be carried up or down stairs. On the second floor, there is a bathroom

Mingle House

> in which there is a bathtub of the first order, having a stop-cock for cold, and another for warm water, so that the person bathing can temper the water to his fancy.

A second, smaller fire damaged the newly rebuilt hotel in 1875 and took the life of Jesse Gantt's wife Caroline. Jesse died in 1880. Under the proprietorship of David Burd from 1882 to 1893, it was called the Central Hotel. David Mingle bought it in 1899, and it became the Mingle Hotel. In 1922, the property passed to Myrtle and Mildred Mingle, and their heir and nephew David M. Myers remodeled the hotel in 1958-59. From 1963 to 1972 Kermit and Arlene Harry owned and operated it as the Newporter Hotel and Towpath Dining. It was restored and remodeled by Jack Gochenour in 1982 as the Newport Hotel and Tavern.

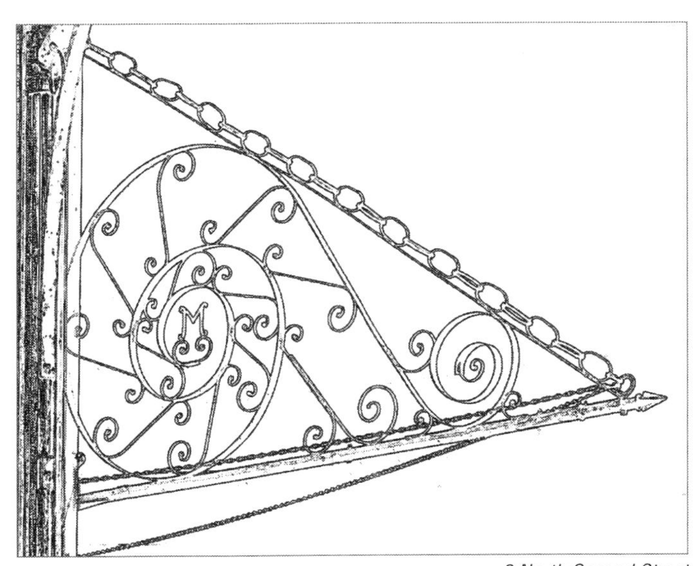

2 North Second Street

14

15 North Second Street

Temple Front

[A] **Pediment**—a triangular section placed above the horizontal superstructure which rests immediately upon the columns. Pediment is a fancier term for gable.

[B] **Cornice**—a roof eave finished with a decorative molding. Eaves, or roof overhangs, are functional. Their purpose is to throw rainwater away from a building's walls and foundation and to shade the building from sunlight during the summer.

[C] **Frieze**—a broad horizontal band of sculpted or painted decoration.

[D] **Dentil**—a small block used as a repeating ornament in a cornice. Dentils are found in classical Greek and Roman architecture and in later eclectic styles.

[E] **Half-column**—an attached column projecting a little more than half its diameter from the wall and serving a structural purpose.

CLASSICAL FAÇADE

The architectural style of this building is Classical Revival, which copied architectural elements of ancient Greek and Roman buildings. It is characterized by symmetrical shape, low roof lines, tall columns, and pediments. It was popular in this country from the early 1800s to the early 1900s. In this bank building, the architect wanted to project a sense of confidence and stability.

FACELIFT IN 1923

From about 1889 to 1891, this property was part of the A. B. Demaree Boot and Shoe Store. The First National Bank, which began

Pre-1923

business in 1875 as the People's Bank of Newport, was reorganized and renamed in 1893 when it constructed this building of brick with arched windows. The original façade was replaced by the Classical Revival white marble in 1923. Portions of the building have been home to Western Union and Cloyd Adams Barbershop. The United Telephone Company Switchboard and private living quarters were on the second floor. In 2018, this building is the home of Orrstown Bank's Newport Square Location.

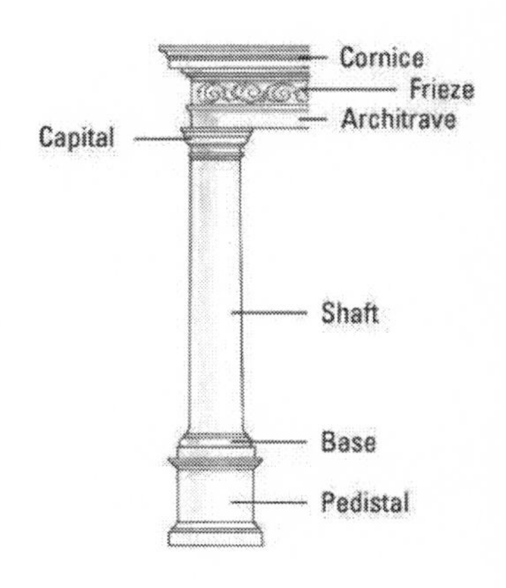

15 North Second Street

1 North Second Street

ANCHORING THE SQUARE

Lintel—a structural horizontal block that spans the space or opening above doors, windows or fireplaces. Lintels can have a structural purpose, can be merely a decorative element, or can be both.

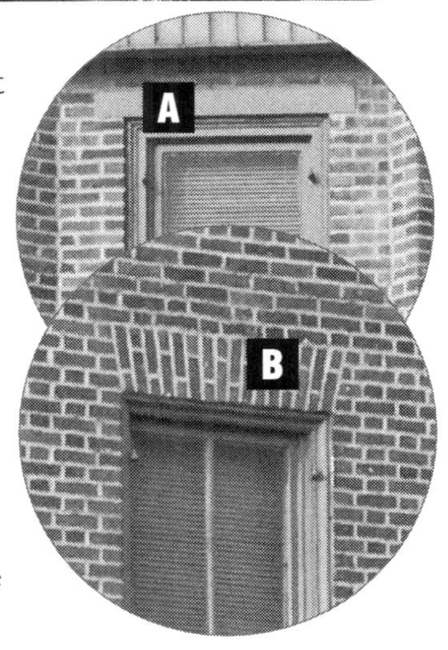

[A] **Horizontal Sandstone Lintel**

[B] **Swayed Brick Lintel**

HIPPED ROOF

This three-story building was constructed of brick on a stone foundation in the mid-1800s.

It is easy to assume the third floor is a later addition as the brickwork pattern and window architecture change. The second-story windows have swayed brick lintels while the third-story lintels are horizontal sandstone.

Notice the widow's walk on the hipped roof in the historical picture.

THE HISTORY

In the mid-19th century, Samuel Huggins operated a hotel and store here, known as the Huggins House. He was elected county sheriff in 1850. His daughter Mary Ann married Benjamin F. Miller, who became sheriff in 1859. Samuel died in 1861 and the hotel subsequently became known as the Miller Hotel. Upon Miller's death in 1892, the building became the Graham Hotel (1894-1905), the Citizens National Bank and the Cumberland County National Bank (CCNB). On the second floor, May Wagonseller had a millinery shop and Frank Milligan and Arthur C. Landis had offices. On the third floor were the Calumet Club and the PEK Fraternity.

Graham Hotel

Early occupants included Jean Boden Beauty Salon, Paul Wilson Barbershop, N. M. Eyth Barbershop, and W. H. Bosserman Boots and Shoes. In the mid-20th century, Alton Comp had a barbershop here and was a representative of the State Capital Savings Association, which was founded by the Pennsylvania Railroad in 1897. He had a small table in the back room of his barbershop where customers would bring their deposits on Friday and Saturday nights. He marked their books and took home a cigar box full of money. On Monday morning, he took the box of money to the main office of the Savings Association. He could moonlight this job by having his wife keep his evening appointments while he traveled around Perry County picking up money for deposit.

In 2018, PNC Bank occupies most of the building while Comp's Barber Shop and Beauty Salon operates on the first floor beyond the bank.

1 North Second Street

16

56 South Second Street

BUILT CIRCA 1910

[A] **Dormer**—a window that projects vertically from a sloping roof. Notice the stone mosaic decoration in the gable of the dormer.

[B] **Lintel**—a structural horizontal block that spans the space or opening in doors, windows, and fireplaces. Lintels can have a structural purpose, can be merely a decorative element, or can be both.

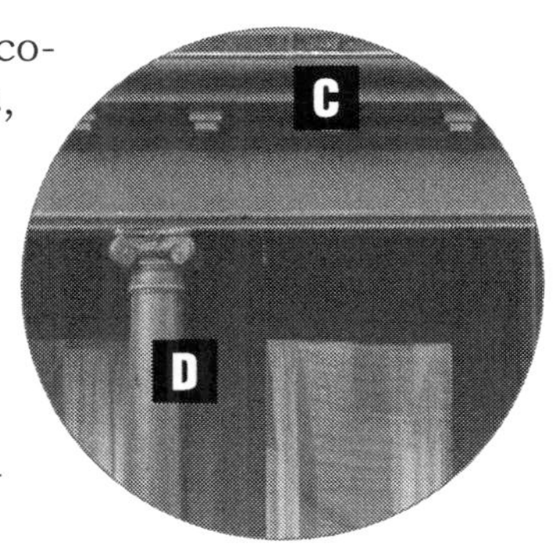

[C] **Cornice**—a roof eave finished with a decorative molding. Eaves, or roof overhangs, are functional. Their purpose is to throw rainwater away from a building's walls and foundation and to shade the building from sunlight during the summer.

[D] **Column**—an upright support. The top of a column, or capital, comes in numerous styles. This one is called Ionic.

ARTS & CRAFTS

Arts & Crafts architecture is simpler than its numerous Victorian predecessors. It is not trying to overwhelm but allows its structural form to reveal its own beauty. Notice the prominent dentils, stained glass in second-story center window, the bay window on the first story, the dormers with mosaic in the gable, and a large one-story full front porch with dentils supported by Ionic columns.

When this two-and-a-half-story house was built around 1910, the architect used brown Roman bricks which are longer and flatter than regular bricks and thin mortar joints called "butter joints" on the front of the house. Out of sight on the side and back of the house are the softer, locally-produced red bricks.

The façade of this house is balanced, not symmetrical.

- Houses that are symmetrical are built with the door in the middle of the façade and the same number of windows on either side.
- Houses that are balanced are built with architectural elements that have similar proportions on subsequent stories. Note that the bay window on the left side of the first story is echoed by the triple window above it on the second story and a wide dormer. To the right of the door, note the single window on the first floor, the single window directly above it and the narrower dormer above them.

A DENTAL PRACTICE

In 1831, Dr. John H. Doling built a small house on the property. Dr. Doling taught school for the 1832-33 term in a building known as "the Barracks", located on the north side of Market Street between Sipe's Tavern (now the Senior Center, on the Square) and the river.

Dr. J. Henry (Harry) McCulloch graduated from Dental School in 1901. When he bought this property in 1910, it contained a two-story frame house and a stable. He tore down the old house and replaced it with this stylish one for his dental practice. By 1920, he and his wife Sara were also living in the house. Dr. McCulloch retired from his practice in 1936 and died in his home in 1951.

56 South Second Street

17

102 South Second Street

BUILT CIRCA 1870

[A] **Drip Cap**—the horizontal molding above the top edge of lintels, windows, and doors. The purpose of a drip cap is to direct water away from the window and directly to the ground.

[B] **Sash Window**—made up of one or more movable "sashes" that form a frame to hold panes of glass, which are often separated from other panes by glazing bars known as mullions. The upper sashes of these windows hold two panes.

[C] **Double Hung Windows**—two sashes that open independently of each other

[D] **Fanlight**

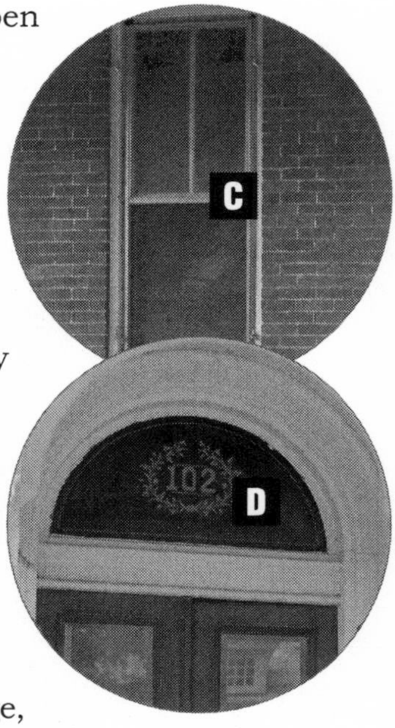

Italianate

This Italianate house shows the typical low-pitched roof, which in this case is also hipped. The projecting eaves probably had brackets at one time, while the pedimented windows retain their elaborate lintels with brackets. The huge Roman-arched doorway hood is supported by square pilasters with three-dimensional detail. Notice the etched house number in the fanlight. Newport acquired its current street names and house numbers in 1894. The double-leaf doors contain large, single panes of glass just becoming available in the late 19th century. The shutters are original and the remains of the original two-over-two windows are displayed in the upper sashes. The floor plan is asymmetrical. Also notice the square, tapered columns supporting the porch.

Snap-on Tools

Very early, this location was the home and business of "Hatter" Smith who made stovepipe hats. Benjamin Himes constructed a brick house here between 1863 and 1876. In 1877, he sold it to the Rev. S. W. Seibert. During the 1920s and 1930s, this house was the home of E. William Myers, president of Newport Forged Steel Products Company from 1920 until his death in 1939. By 1931 Myers had become the president of Snap-on Tools of Kenosha, Wisconsin as well. The two companies merged in 1945 under the name Snap-on Tools. After Myers' death, the company continued to operate in Newport until 1955. Myers was also the largest shareholder in Remington Rand. He hoped to establish plants for both Snap-on Tools and Remington Rand in Oliver Township near the Big Buffalo Creek, but he died while traveling by train from Syracuse, New York to Milwaukee, Wisconsin to negotiate the deal.

102 South Second Street

18

141 South Second Street

One of Its Kind

[A] Tower—an architectural element that is significantly taller than it is wide and that originates at ground level.

[B] Bracket—a structural or decorative element that supports an eave or cornice. Although brackets appear to support the eave or overhang, their presence may be merely decorative. Brackets can be made of wood, stone, plaster, or metal.

[C] **Corbel**—projection of brick jutting out from a wall to support a structure above it.

[D] **Scrollwork**—decoration consisting of spiral lines or patterns, especially as cut by a scroll saw.

QUEEN ANNE

The house is solidly Queen Anne with its prominent two-story bay windows with corner brackets and a steeply pitched roof, a central tower with a steep roof, and a side porch with fluted Doric columns which give the house an asymmetrical feel. Notice how the placement of the eave brackets and the contrasting paint colors accentuate the verticality of the house.

KEEP IT IN THE FAMILY

George Fleisher and his brother John were the contractors responsible for building the 1866 Newport School on North Fourth Street. George was one of the first members of St. Paul's Lutheran church and helped in the construction of the church. He was director of the First National Bank of Newport and one of the first trustees of the Newport Cemetery Association.

In 1885, George bought out the other owners of the Newport Manufacturing and Building Company to become the sole owner of what became known as the Newport Planing Mill Company. According to H. H. Hain, he "achieved his boyish ambition to become the owner of a planing mill. He erected many homes in Newport during his lifetime and helped make the town what it is." In 1900, he sold the mill to his son J. Emery who continued to operate it until 1929.

George acquired this land from his brother John in 1870. He built this house and lived here with his wife Mary E. (Long) Fleisher. Upon George's death in 1917, the house passed to his daughter Catherine E. Fleisher, and she conveyed it to her niece Eleanor F. Shutter in 1926.

141 South Second Street

19

67 South Second Street

DAYDREAM SALON AND SPA

[A] Queen Anne Window—a window with small lights or panes arranged in various patterns. Usually found in the top sash only.

[B] Turret—a small circular tower that projects vertically from the wall of a building. Turrets typically do not start at the ground level. Instead, they cantilever out from another upper level. The size of a turret is therefore limited by technology since it puts additional stresses on the structure of the building.

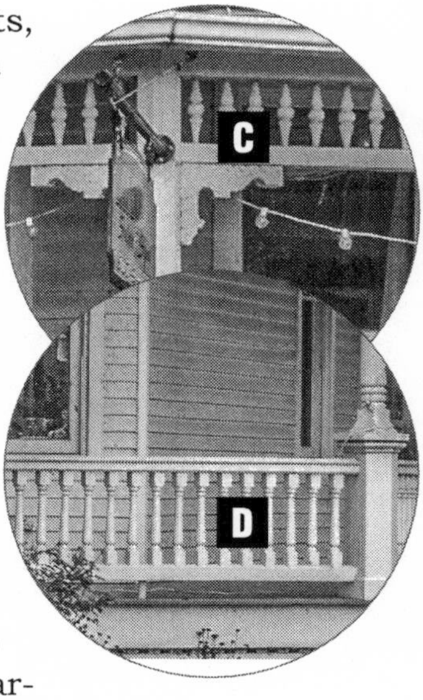

[C] **Spindlework**—delicately turned supports, used as a frieze suspended from the porch ceiling.

[D] **Baluster**—a short pillar or column, typically decorative in design, placed in a series to support a handrail. Balusters have several purposes within the overall structure: support the handrail, provide safety, and provide additional style and structural flair.

C

D

Queen Anne

The architectural style of this house was popular in the United States from 1880 to 1910. Queen Anne architecture was characterized by many elements that can be found in this house: wrap-around porch; an asymmetrical façade; differing wall textures; delicately turned porch posts; lacy, ornamental spindles; bay windows; monumental chimneys; turrets, painted balusters; and wooden or slate roofs. Queen Anne houses were usually frame buildings [not brick or stone] which made it easier to get away from a rectangular building.

Notice the balusters have been turned on a lathe, a more expensive process than a simple cut with a saw. The varying scalloped and rounded shingles on the turret create additional textures. The dormers are of different styles. The windows are narrow and tall, which creates a vertical effect.

From Bar to Spa

John Fite built the first house on this property in 1831, where he operated a tavern. George Hummel was living here in 1872 when he sold the property to Jacob Tibbens. In 1884, Tibbens sold it to William H. Gantt, who built the house we see today. Samuel Adams Sharon bought the house in 1907. He operated a lumber yard on Third Street selling railroad ties and owned the property in Oliver and Miller townships which became Sharon Orchards. In 2018, this property is the Daydreams Salon and Spa.

67 South Second Street

20

51 South Second Street

BUILT CIRCA 1890

[A] Frieze—a broad horizontal band of sculpted or painted decoration.

[B] Drip Cap—the horizontal molding above the top edge of lintels, windows, and doors. The purpose of a drip cap is to direct water away from the window and directly to the ground.

[C] Bracket—a structural or decorative element that supports an eave or cornice.

Although brackets appear to support an eave or overhang, their presence may be merely decorative. Brackets can be made of wood, stone, plaster, or metal.

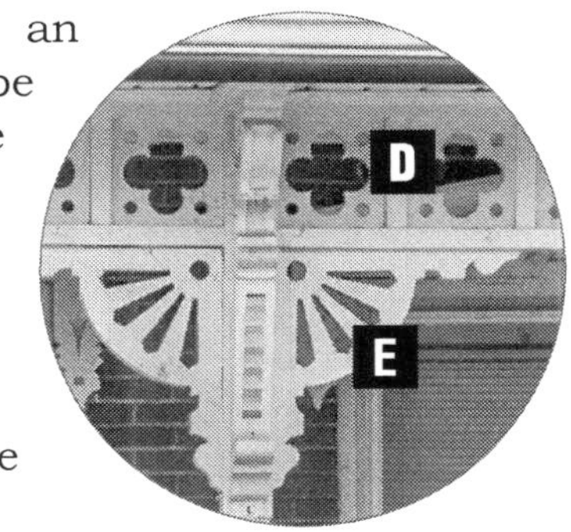

[D] **Quatrefoil**—a type of decorative framework consisting of a symmetrical shape which forms the overall outline of four partially overlapping circles of the same diameter.

[E] **Sunburst**—an architectural ornament that consists of rays or "beams" radiating out from a central disk to evoke sunbeams. Sunbursts can be circular, semicircular or semi-elliptical in shape.

PICTURESQUE

The Picturesque movement was a departure from the formality of the Georgian period, reflecting a romantic desire to return to nature and influencing architecture throughout the Victorian period. This two-story brick house has five bays and a modern storefront on the north side. Elaborate scrollwork adorns the drip caps and frieze board. There are eave brackets and a two-thirds front porch highlighted by gingerbread and turned wooden columns with scrollwork brackets. The ornamentation includes stylized versions of natural floral elements and allows an interesting interplay of light and shadow, while the porch draws people out of doors.

The Schlomers with their harness

A musical interlude

The industrial revolution made such details possible. Many of these elements were machine made. The materials are elaborate but not beyond the means of the middle class. This sort of detail can be found in small towns all around the country because they could be mass produced via steam scroll saw and shipped cheaply by railroad. The Newport Planing Mill, which excelled in such fine scrollwork, was just around the corner.

CLOTHIER

The Kough brothers acquired this lot from A. B. Demaree in 1867. Their warehouse, which stored and sold grain and agricultural products, was in the alley behind. In 1889, they sold the lot to Peter Schlomer, a 28-year old immigrant from Germany, for $800. Peter Schlomer first operated a harness business, and in 1896, he expanded into selling clothing. Schlomer sold the property with its current house to P. K. Brandt in 1905 for $5000. It has since been Charles Brandt Clothing, Bernie Carl Clothing, and Charles Welfley Clothing.

51 South Second Street

References

Beach, Nichols. "Oliver Township, Newport, Juniata River." *Atlas of Perry, Juniata and Mifflin Counties, Pennsylvania*. Philadelphia, Pennsylvania: Pomeroy Whitman & Co. 1877. Digital image. *Historic Map Works, Rare Historic Maps Collection*, http://www.historicmapworks.com/Map/US/160120/Oliver+Township++Newport++Juniata+River/Perry+County+1877/Pennsylvania/.

Becker, Margie. *A Scrapbook of Schoolhouses in Perry County: Buffalo, Howe, Juniata, Miller, Newport Borough & Oliver Townships*. Newport, Pennsylvania: The Perry Historians. 1997

"Deeds". *Landex Remote*. Digital images. https://www.landex.com/land-records-access.asp .

Dinsmore, Douglas. *National Register of Historic Places Inventory Nomination Form: Newport Historic District*. Print. National Park Service, National Register of Historic Places. Washington, D.C., 1998.

Fowler, T. M. and James B. Moyer. *Newport, Perry County, Pennsylvania*. Morrisville. *1895*. Digital image. *Library of Congress Geography and Map Division Washington, D.C.* https://catalog.loc.gov/vwebv/search?searchCode=STNO&searchArg=83693865&searchType=1&recCount=10.

Hain, Harrison. H. *History of Perry County, Pennsylvania, including descriptions of Indians and pioneer life from the time of earliest settlement*. Harrisburg, Pennsylvania: Hain-Moore Company. 1922.

Hartzell, Michael. "Newport from 1829 Down to the Present." *The News* (Newport) July 28, July 31, August 14 and August 28, 1880. Digital images. *Newspapers.com*. https://www.newspapers.com: 2017.

Heberling, Scott D. *Historic Resources Survey Report, Canal-Era Resources in Mifflintown Borough, Juniata County and Millerstown and Newport Boroughs, Perry County, Pennsylvania*. Heberling Associates, Inc. Alexandria, Pennsylvania in association with Allegheny Ridge Corporation. Altoona, Pennsylvania: 2017.

Historical Society of Perry County: Subject matter and family files.

"Historic Walking Tour". *Newport Pennsylvania, Preserving our Future*. Newport Revitalization, Inc. no date.

Hopkins, G. M. and C. E. Smith. *Map #123 - Map of Juniata, Mifflin, and Perry Counties, Pennsylvania*, 1863. Philadelphia, Pennsylvania: Gallup & Hewitt, Publishers, 1863. Digital image. *Pennsylvania Historical and Museum Commission, State Archives*, http://www.phmc.state.pa.us/bah/dam/mg/di/m011/Map0123Interface.html.

Long, Rebecca Loy. "The Early Days of Newport". Unpublished manuscript. No date.

Memories of Newport residents: Edna H. and Elmer R. "Bob" Baker, Jr., Suzanne B. Beamer, Jerry A. Clouse, Robert M. Flickinger, Harry A. Focht, Jean H. Tuzinski and Jeffery Lynn Wright.

"Newport 1840-1990". *The Perry Review*. vol. 15. The Perry Historians. 1990.

Newport Centennial Celebration, 1840-1940. No publisher. No date.

"Newport from 1829 Down to the Present". *The Perry Review*. vol. 4. The Perry Historians. 1979.

Newport Revitalization and Preservation Society, Louise Beard Memorial: Photographs and family records.

Newport Sesquicentennial Self-Guided Walking Tour. Unpublished manuscript. 1990.

The News (Newport). Digital images. *Newspapers.com*. https://www.newspapers.com: 2017.

The Perry Historians Library Holdings: Deeds, estate files, wills, subject matter files and genealogical files.

Sanborn Fire Insurance Maps 1884, 1891, 1896, 1902, 1911, 1925. New York, New York: Sanborn Map Company. Digital images. Penn State University Libraries, https://collection1.libraries.psu.edu/cdm/search/collection/maps1/searchterm/newport%20perry%20county%20pa/field/all/mode/all/conn/and/order/nosort/ad/asc/cosuppress/1

A Sesquicentennial Commemorative Book for Newport, PA. 1840-1990. No publisher. No date.

About the Contributors

IRENE S. VANBUSKIRK

Irene began photographing and sharing her work in 2013. Her horses, birds, and countryside around her Perry County home were the initial inspirations for her photography, and the landscapes of Perry County continue to be a thread in her collections. She is also attracted to the geometry and patterns of urban landscapes as well as capturing impressionistic and abstract images. Small towns in the central Pennsylvania area have been frequent locations for her photography.

Irene S. VanBuskirk

In 2014 she attended "In Focus," a Street Photography workshop in New York City, which influenced her to to include people in her photography. Rather than portraiture, most of the people in her photographs represent the human element in the environment.

Irene's work is displayed at the PCCA Gallery, PCCA Art on Tour venues, The Gallery At Second in Harrisburg, CALC in Carlisle, and the Art Association of Harrisburg Community venues. She has also shown her work in juried shows in Philadelphia and Bethlehem. She was the first prize winner in the most recent Camp Hill Plein Air photography competitions - 2014 and 2015.

Although Irene does some traveling she most enjoys capturing artful elements in everyday life, prompting others to see designs and beauty that they might have otherwise missed.

JANE T. HOOVER

Jane's curiosity about local history was awakened early when her family stopped to read every highway historical marker while taking

Sunday afternoon drives. It continued as she rode her horse around the countryside, taking advantage of every farm track and abandoned roadway that she could find. Much later, when she was living in southeastern Schuylkill County and Midtown Harrisburg, she traced their histories as well. When she came to Newport in 2009 it was natural for her to become interested in Newport history. As preparation for Perry County's 2020 bicentennial got underway, she began research for the historical markers and soon became Editorial Director for the Perry County Bicentennial Committee.

Jane T. Hoover

Jane received a B.A. from the University of Pittsburgh, majoring in political science and history. She acquired her editorial skills while working for the Commonwealth of Pennsylvania. Her supervisor had what must have been the sharpest red pencil in Harrisburg. In 1986, she bargained dressage lessons in exchange for editing a book her trainer wished to publish. The book, entitled *Beyond the Mirrors*, was published in 1988 and is still available online.

In addition to local history, Jane is involved with her church, particularly in spiritual formation and environmental stewardship. She and her cat live in Newport where she can easily walk around town and appreciate its history and beauty.

ROGER A. SMITH

When he was in ninth grade, Rog Smith reported to his guidance counselor that he wanted to become an architect. It didn't happen. However, as the decades-long co-owner of a farm and summer camp, he scratched that creative itch by designing and constructing numerous buildings. Rog spent the last eight years of his professional life serving as the Executive Director of the Perry County Council of the Arts. It was there that he had the privilege of overseeing the transformation of Landis House in

Roger A. Smith

Newport from a home suffering from years of deferred maintenance into a community treasure.

Also during this time, Rog developed an appreciation for the architecture of downtown Newport. As a member of the Perry County Economic Development Authority and the Perry County Bicentennial Committee [2020], he determined that an exhibition calling attention to the architecture of Newport's Victorian era could appeal to numerous audiences. With this kernel of an idea, he was gratified when Irene VanBuskirk and Jane Hoover enthusiastically jumped on the bandwagon, bringing their own skills and vision to the collaboration.

This book is an outgrowth of the exhibition called Downtown Details, which opened at Landis House in May 2018 and was subsequently shown at the Veterans Memorial Building in New Bloomfield and Newport High School.

Made in the USA
Columbia, SC
25 April 2018